AF379992

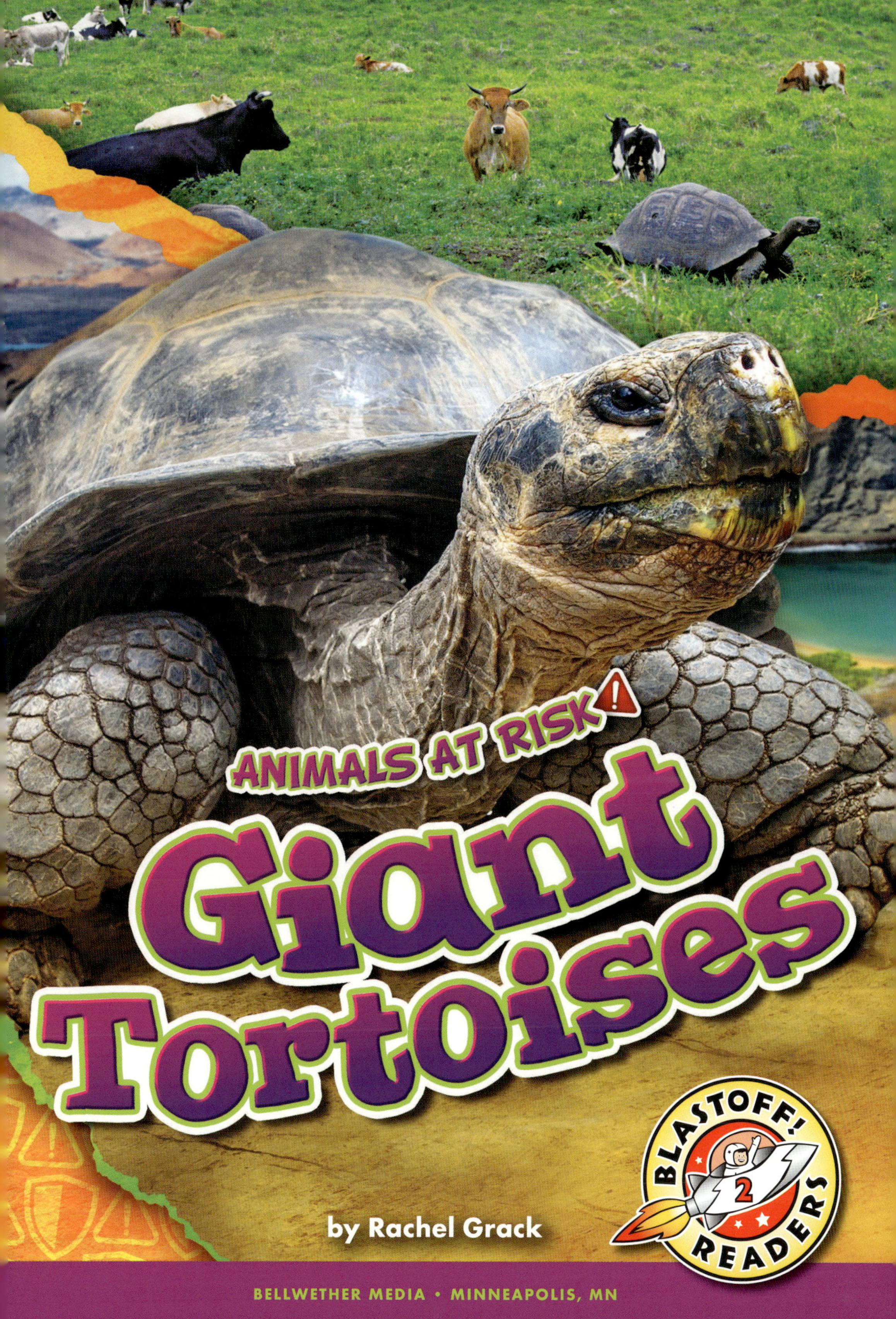

ANIMALS AT RISK
Giant Tortoises
by Rachel Grack
BLASTOFF! READERS
2
BELLWETHER MEDIA • MINNEAPOLIS, MN

Blastoff! Readers are carefully developed by literacy experts to build reading stamina and move students toward fluency by combining standards-based content with developmentally appropriate text.

Level 1 provides the most support through repetition of high-frequency words, light text, predictable sentence patterns, and strong visual support.

Level 2 offers early readers a bit more challenge through varied sentences, increased text load, and text-supportive special features.

Level 3 advances early-fluent readers toward fluency through increased text load, less reliance on photos, advancing concepts, longer sentences, and more complex special features.

★ **Blastoff! Universe**

Reading Level

Grade
K

Grades
1–3

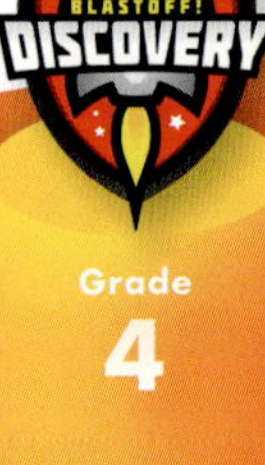
Grade
4

This edition first published in 2023 by Bellwether Media, Inc.

Library of Congress Cataloging-in-Publication Data

Names: Koestler-Grack, Rachel A., 1973- author.
Title: Giant tortoises / Rachel Grack.
Description: Minneapolis, MN : Bellwether Media, 2023. | Series: Blastoff! readers: animals at risk | Includes bibliographical references and index. | Audience: Ages 5-8 | Audience: Grades 2-3 | Summary: "Relevant images match informative text in this introduction to giant tortoises. Intended for students in kindergarten through third grade"-- Provided by publisher.
Identifiers: LCCN 2022000413 (print) | LCCN 2022000414 (ebook) | ISBN 9781644877128 (library binding) | ISBN 9781648347580 (ebook)
Subjects: LCSH: Galapagos tortoise--Juvenile literature. | Testudinidae--Juvenile literature. | Galapagos tortoise--Conservation--Juvenile literature.
Classification: LCC QL666.C584 K64 2023 (print) | LCC QL666.C584 (ebook) | DDC 597.92/46--dc23/eng/20220113
LC record available at https://lccn.loc.gov/2022000413
LC ebook record available at https://lccn.loc.gov/2022000414

Editor: Kieran Downs   Designer: Brittany McIntosh

Printed in the United States of America, North Mankato, MN.

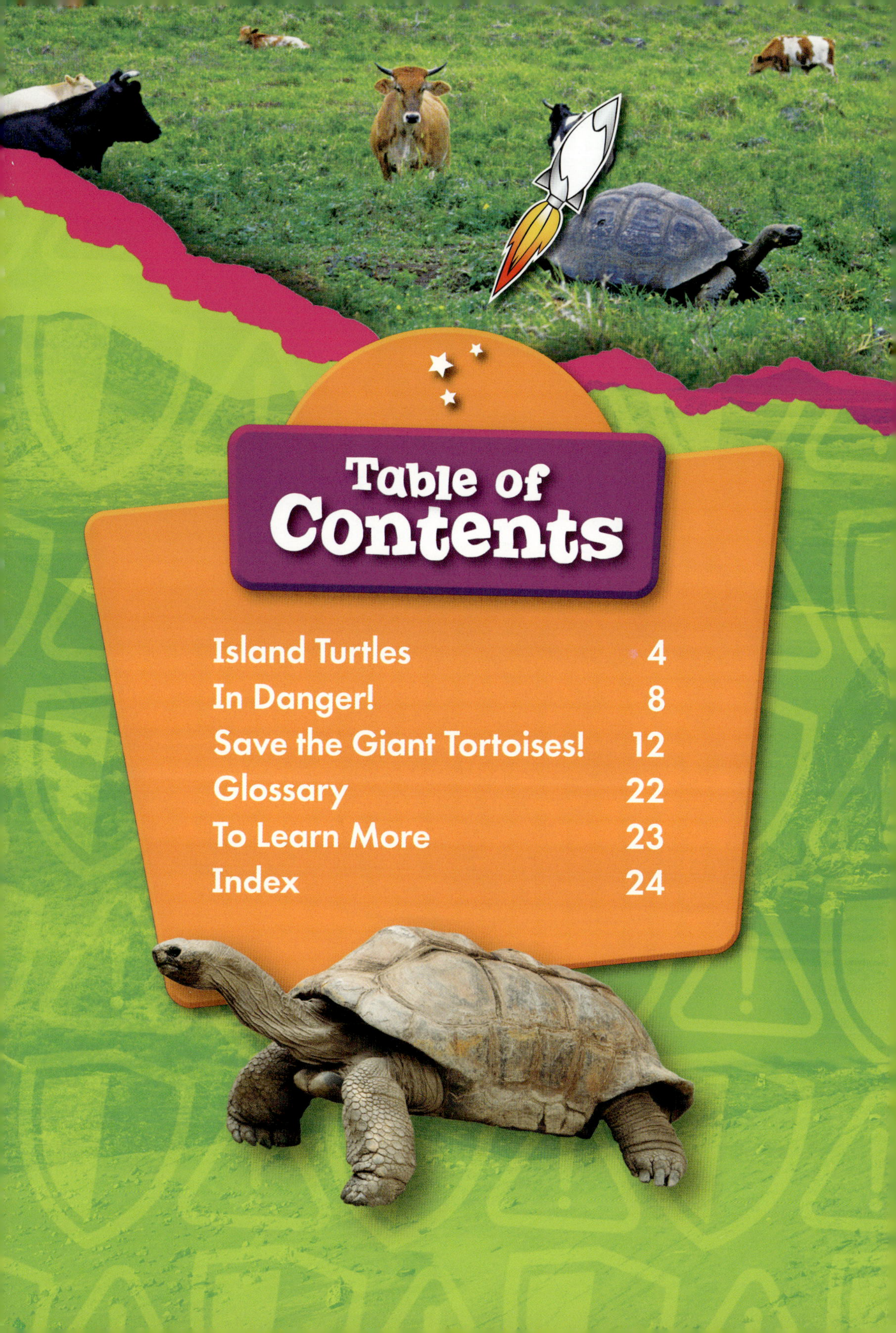

Table of Contents

Island Turtles

Giant tortoises are huge
land turtles. Many **species** live
throughout the Galápagos Islands.

4

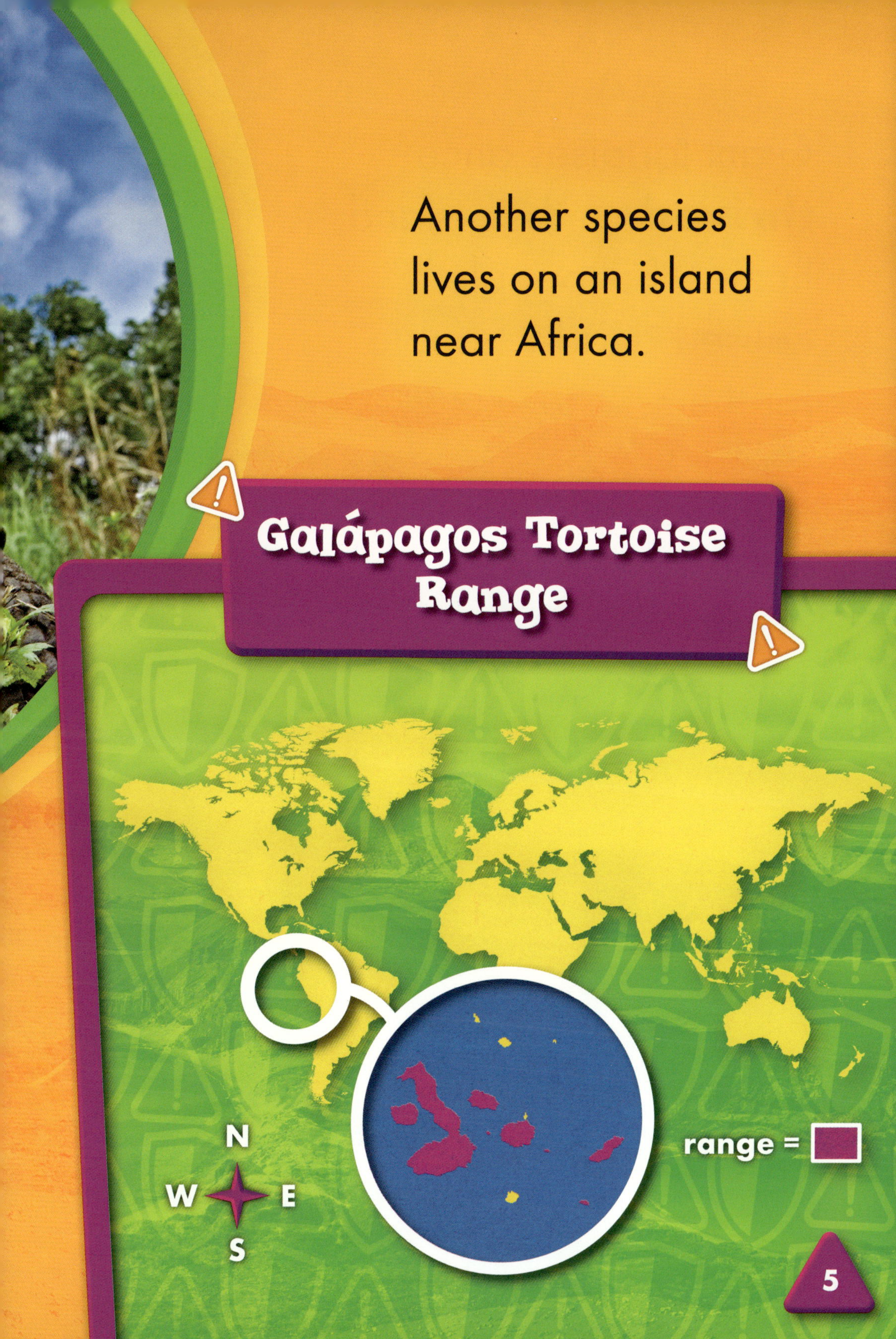
Another species lives on an island near Africa.
Galápagos Tortoise Range
N
W
E
S
range =
5

Giant tortoises once **thrived** on their island homes. But people caused deadly trouble for them.

Today, most species are **endangered**. Some have already gone **extinct**.

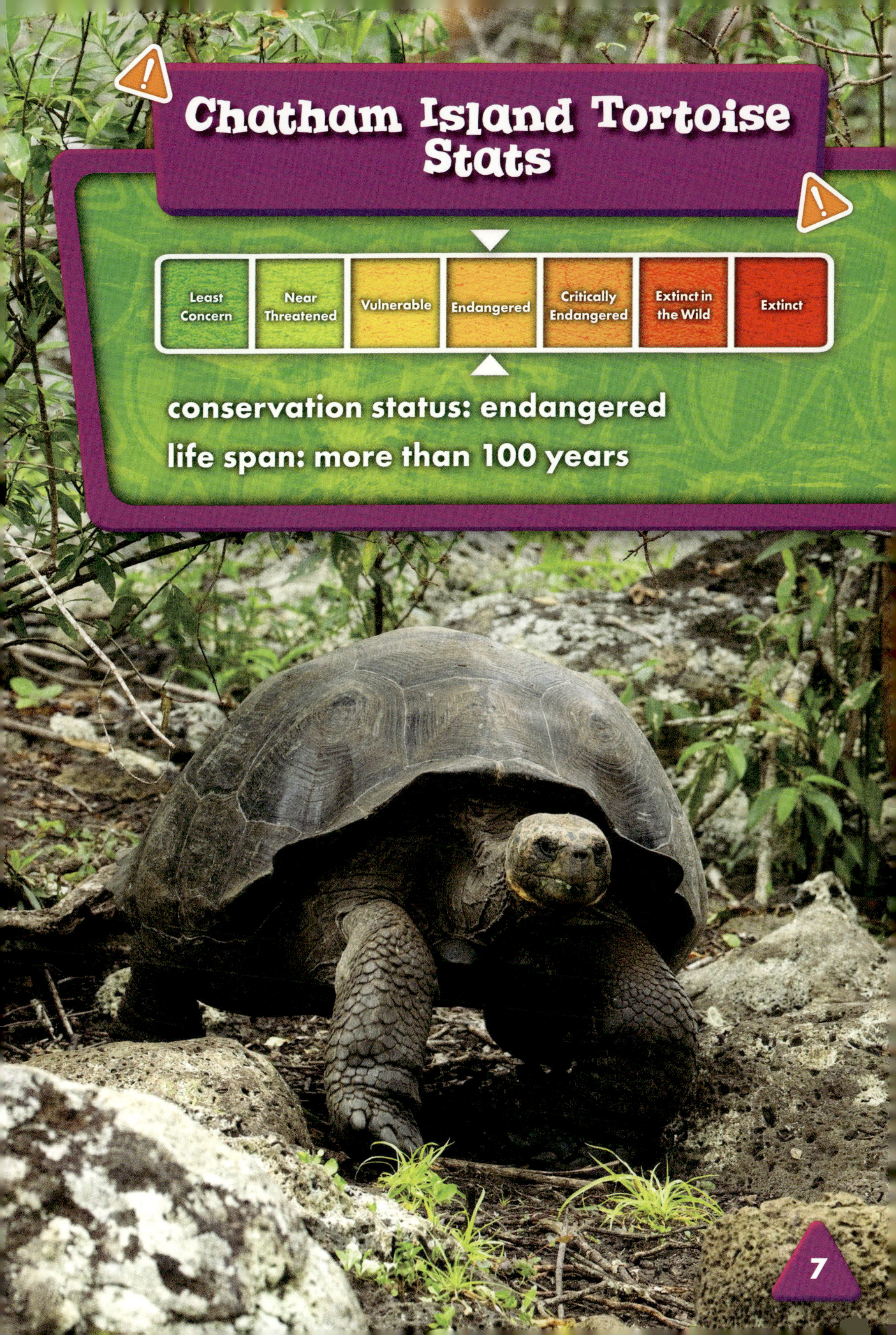

Chatham Island Tortoise Stats

Least Concern
Near Threatened
Vulnerable
Endangered
Critically Endangered
Extinct in the Wild
Extinct

conservation status: endangered
life span: more than 100 years

Long ago, sailors brought
new animals to giant tortoise
habitats. The new animals
ate the tortoises' food.

Some animals ate
the tortoises' eggs
and young.

Threats

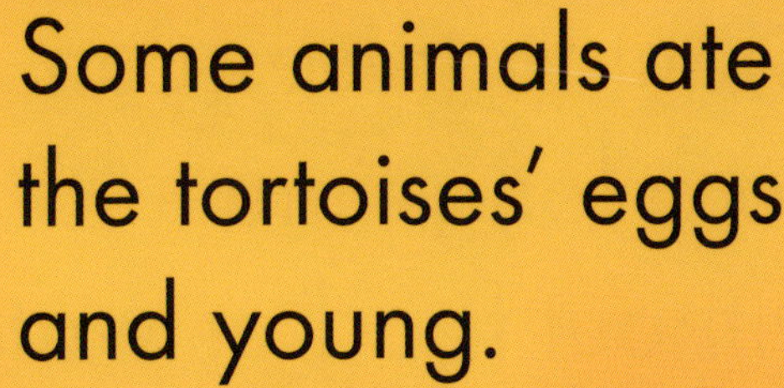

sailors landed
on islands

new animals
left behind

less food and
fewer eggs and young

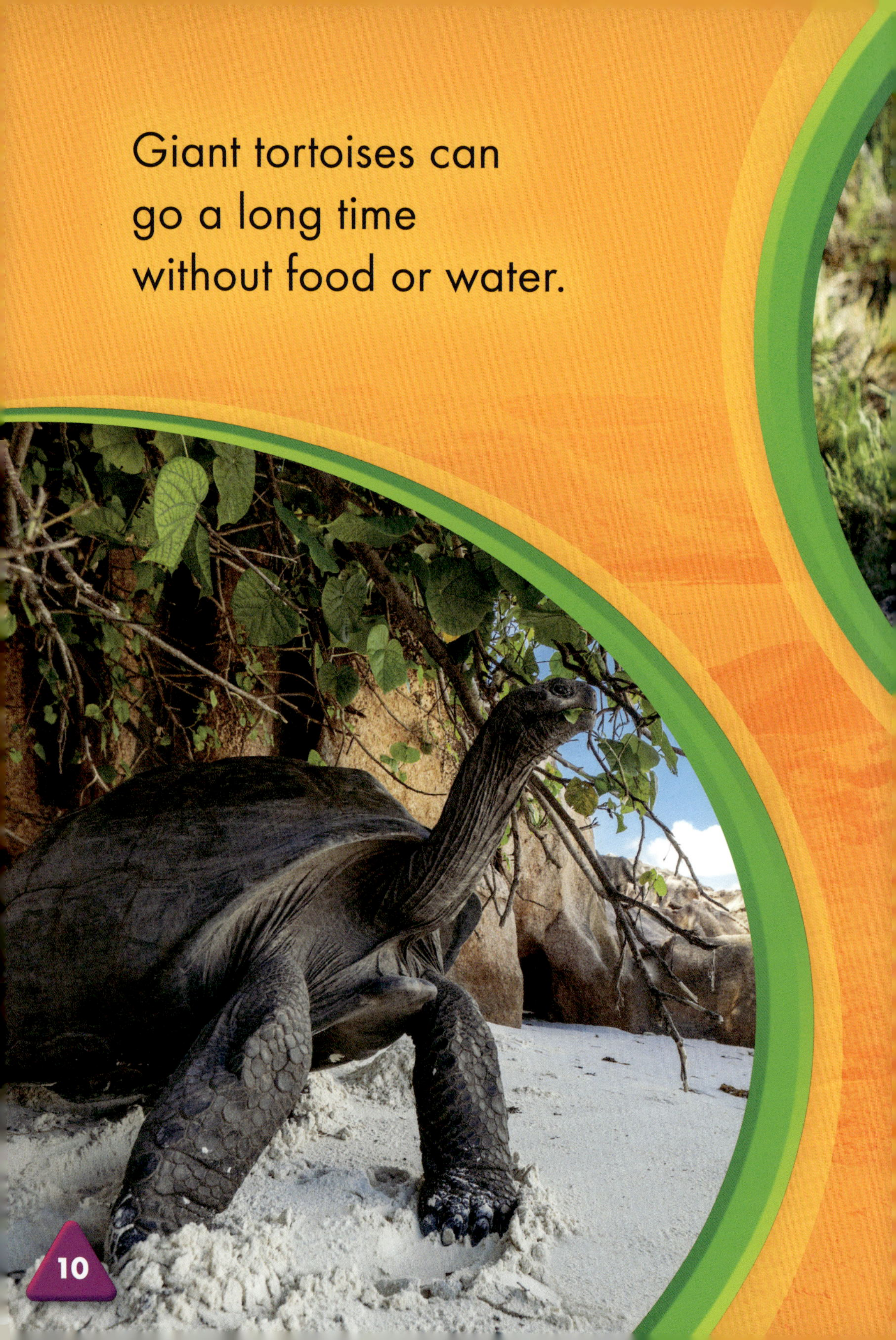

Giant tortoises can
go a long time
without food or water.

They made good animals
for ship travel. Sailors took them
for fresh meat at sea.

Save the Giant Tortoises!

Giant tortoises open pathways for other animals. This makes more room for plants to grow.

They also spread seeds. Without them, their island **ecosystems** would suffer.

The World with Giant Tortoises

1 more giant tortoises

2 seeds spread

3 healthy island ecosystems

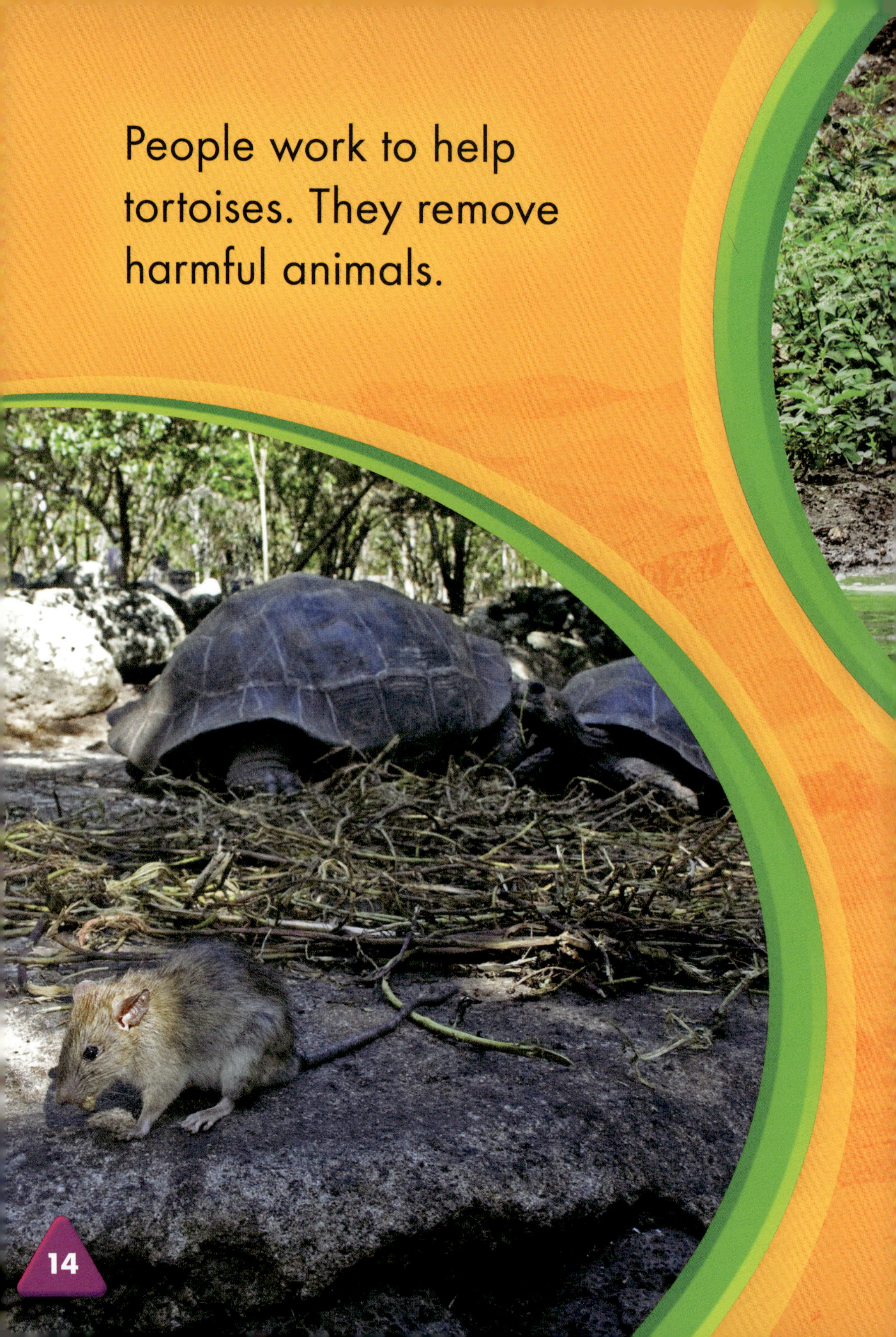

People work to help
tortoises. They remove
harmful animals.

Governments set aside land
for habitats. Tortoises and
their young stay safe.

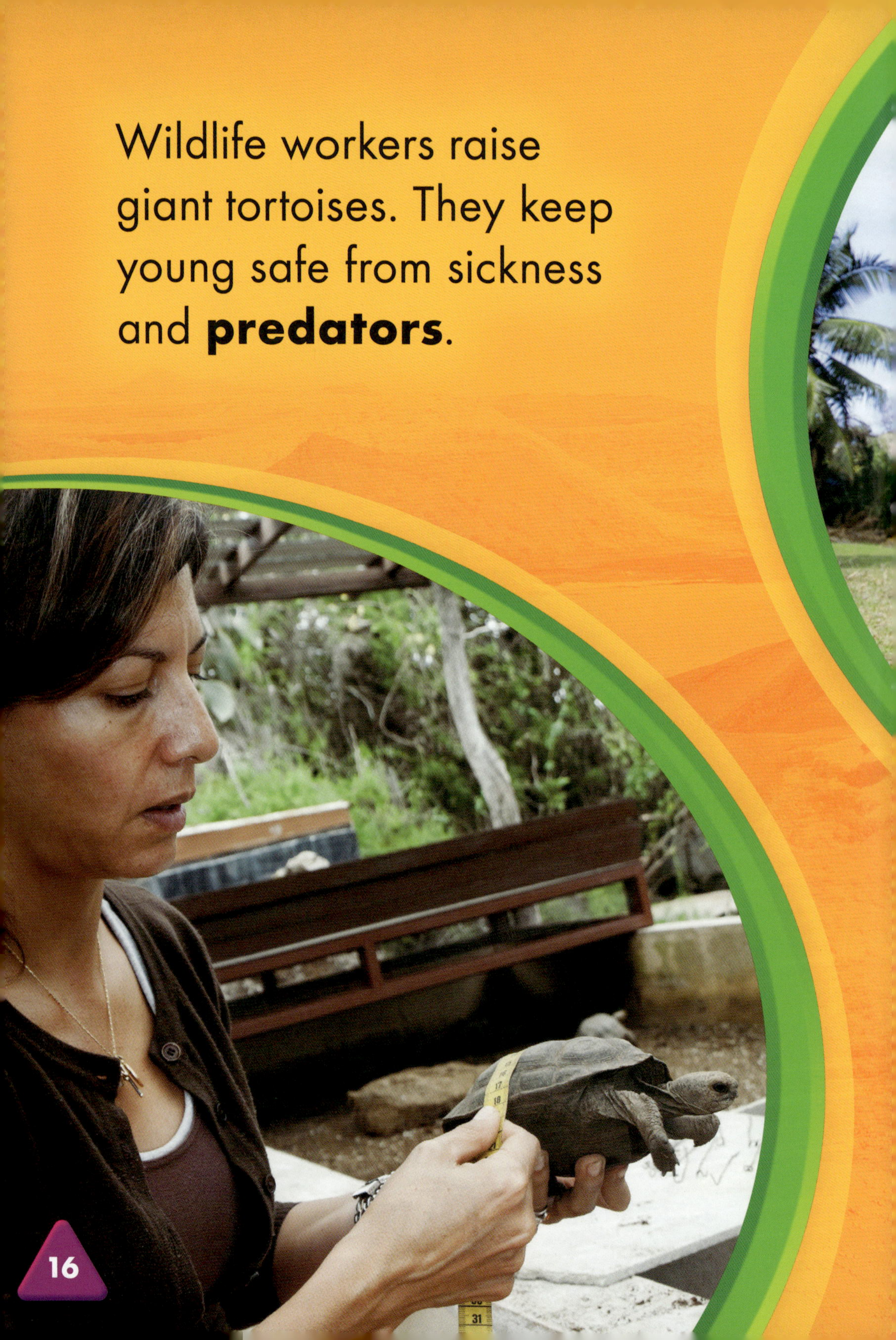
Wildlife workers raise giant tortoises. They keep young safe from sickness and **predators**.

16

Tortoises later return to the wild.

People study giant tortoise **migration**. They work with landowners to make sure paths are clear.

Tortoises can move freely
and live unbothered.

Donations to wildlife groups help giant tortoises. **Adopting** a giant tortoise is another way to give.

Together, everyone can save these grand turtles!

Glossary

adopting—taking over the care for someone or something; people who adopt wild animals give money for someone else to care for them.

donations—gifts for a certain cause; most donations are money.

ecosystems—communities of plants and animals living in certain places

endangered—in danger of dying out

extinct—to have completely died out in the wild

habitats—the places and natural surroundings in which plants or animals live

migration—the act of traveling from one place to another, often with the seasons

predators—animals that hunt other animals for food

species—kinds of animals

thrived—lived well

To Learn More

AT THE LIBRARY

Grack, Rachel. *Sea Turtles*. Minneapolis, Minn.: Bellwether Media, 2022.

Jaycox, Jaclyn. *Giant Tortoises*. North Mankato, Minn.: Pebble, 2021.

Rossiter, Brienna. *Saving Earth's Animals*. Lake Elmo, Minn.: Focus Readers, 2022.

ON THE WEB

Factsurfer.com gives you a safe, fun way to find more information.

1. Go to www.factsurfer.com.

2. Enter "giant tortoises" into the search box and click 🔍.

3. Select your book cover to see a list of related content.

Index

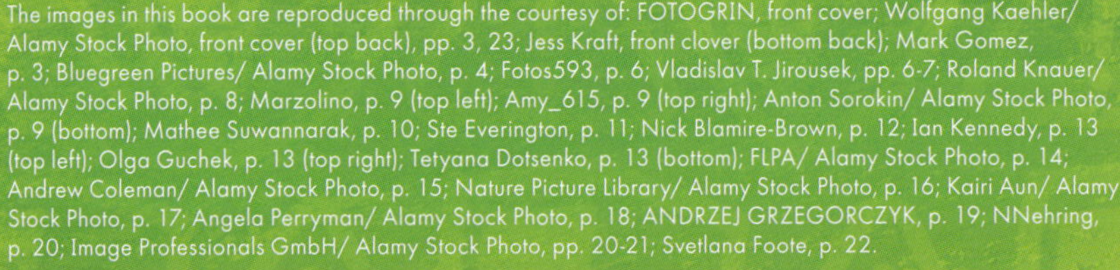

The images in this book are reproduced through the courtesy of: FOTOGRIN, front cover; Wolfgang Kaehler/ Alamy Stock Photo, front cover (top back), pp. 3, 23; Jess Kraft, front clover (bottom back); Mark Gomez, p. 3; Bluegreen Pictures/ Alamy Stock Photo, p. 4; Fotos593, p. 6; Vladislav T. Jirousek, pp. 6-7; Roland Knauer/ Alamy Stock Photo, p. 8; Marzolino, p. 9 (top left); Amy_615, p. 9 (top right); Anton Sorokin/ Alamy Stock Photo, p. 9 (bottom); Mathee Suwannarak, p. 10; Ste Everington, p. 11; Nick Blamire-Brown, p. 12; Ian Kennedy, p. 13 (top left); Olga Guchek, p. 13 (top right); Tetyana Dotsenko, p. 13 (bottom); FLPA/ Alamy Stock Photo, p. 14; Andrew Coleman/ Alamy Stock Photo, p. 15; Nature Picture Library/ Alamy Stock Photo, p. 16; Kairi Aun/ Alamy Stock Photo, p. 17; Angela Perryman/ Alamy Stock Photo, p. 18; ANDRZEJ GRZEGORCZYK, p. 19; NNehring, p. 20; Image Professionals GmbH/ Alamy Stock Photo, pp. 20-21; Svetlana Foote, p. 22.